Heinkel

He162

By
The Aeronautical Staff of Aero Publishers, Inc.

In Cooperation With
Edward T. Maloney, Curator of The Air Museum

Scale Drawings by Uwe Feist

Distributed by:
Airlife Publishing Ltd.
7 St. John's Hill, Shrewsbury SY1 1JE, England

© **AERO PUBLISHERS, INC.**

1965

Library of Congress Catalog Card Number

65-26827

ISBN-0-8168-0512-1

Printed and Published in the United States of America by Aero Publishers, Inc.

HEINKEL 162 DEVELOPMENT

By EDWARD MALONEY
Curator THE AIR MUSEUM
Ontario, California

One of the truly remarkable Jet Fighters to come out of World War II, was the Heinkel 162A "People's Fighter" of 1944.

This unique aircraft was the pride of Dr. Ernst Heinkel who built his first aircraft in 1911. He designed aircraft for the Albatros Works at Johannistal, near Berlin, during World War I. His designs won many competitions and prizes. In 1914, he became Albatros' Chief Designer and created no less than thirty different aircraft models during this period.

In 1922, he founded the Heinkel Aircraft Works at Warnemünde. His company progressed with the rise of the German Luftwaffe. Among his better known creations are — the Heinkel He-70 "Blitz," one of the most aerodynamically clean and fast commercial transports of the early 1930's; the Heinkel He-111 — standard Luftwaffe bomber of World War II 1934-1945; the Heinkel He-100 fighter, which in 1939 established a World Speed record of 463.9 mph; the Heinkel He-112 which became a standard service fighter of the Luftwaffe.

During World War II, it was generally believed in America that the Italian CAPRONI-CAMPINI N-1 was the first jet propelled aircraft to fly. Later it was learned the British Gloster E 28/39 was the world's first pure jet-gas turbine aircraft to fly successfully. In fact, however, the German Heinkel He-178 had flown as early as August 1939, but due to high level secrecy of the German Air Ministry (R.L.M.), this fact was not made known to the world until after the end of World War II.

Heinkel became interested in gas-turbine engines in 1936. He personally financed and built the development of this jet engine—the Heinkel He S-3B. Two years were spent developing this jet turbine, 1936-1938, and it produced a thrust of 1,100 lbs. The development was kept secret from the R.L.M. to prevent interference from outside sources. The R.L.M. upon learning of this new invention were determind to reject it.

The Heinkel He-178 was a shoulder-wing research aircraft which featured a wood wing and all metal fuselage and a retracting landing gear. It had a maximum speed of 435 mph. On August 24, 1939, with Captain Warsitz as Test Pilot, it made its first flight at Rostock-Marienehe Airfield. A demonstration flight was later made for R.L.M. officials. General Udet was enthusiastic but Gen. Milch remained cool towards this new form of jet propulsion.

Heinkel continued improving his jet engine development and his next successful gas turbine was the He S-8A. He then designed the Heinkel He-280 which featured two of these He S-8A engines. The He-280 was the first German jet propelled fighter aircraft design. It made its first flight in April 1940 and attained a speed in excess of 500 mph but still the R.L.M. was unimpressed—their attitude was—"We will win this war with conventional fighters" (Me-109 and FW-190). Note that this attitude prevailed when Hitler was conquering Europe—1940.

The R.L.M. was incapable of appreciating Heinkel's significant development of jet propulsion. His He-178 was placed in a Berlin museum as a curiosity and the eight prototype He-280's were used only as engine test beds. Later, after the death of Ernst Udel—Heinkel's only friend in the Air Ministry, the R.L.M. completely ignored Heinkel's contributions and capabilities and when the significance of the jet aircraft came to light, the R.L.M. Engineering Division, gave the B.M.W. and the Junkers Companies the contracts to design jet engines. The Arado and Messerschmitt companies were given contracts to build jet aircraft.

Heinkel was completely **ostracized** from the competition. This waste of time and talent was disastrous for the Luftwaffe for it was late 1944 before the Arado 234 and Me-262 made their appearance on the war front.

In 1944, the R.L.M. realized that there was a pressing need for a mass produced jet fighter effective in air defense.

The R.L.M. invited five German aircraft firms in the summer of 1944 to submit bids for a simple, high performance jet fighter. Although the war was going very badly for Germany, Heinkel accepted the invitation for several reasons. First, it was a challenge to his engineering background and second he wanted the recognition due him, having been ostracized twice from previous design competitions.

In 1944, there arose the cry for a "Volksjaeger" or People's Fighter. It had to be built of non-strategic materials and in a very short time and had to be easy to fly. It was further specified that the design, development, and construction of the prototypes and pre-production models had to be completed to the point of mass production readiness within six months. This was quite an order. Such a proposal would be unheard of even today.

The "People's Fighter" specifications were finally issued in September 8, 1944: Top speed 466 mph with a B.M.W. .003 jet engine of 1,760 lbs. thrust; armament one or two cannon; endurance 20 minutes at low altitude; gross weight not to exceed 4,400 lbs. and be able to take off within 445 yards. Deadline—be ready for mass production by January 1, 1945.

Herr. Heinkel designed a jet aircraft of simplicity. It was simple to construct, had economy of materials and ease of maintenance. He designed the jet engine atop the fuselage, as there was not time to run tests on intake inlets and exhaust exits as would have been done had the engine been mounted within the fuselage, and also the engine would not be damaged in the event of a belly landing.

The safety of the pilot was not overlooked and so Heinkel installed a pilot-ejection seat. It was the first production fighter to do so. It was a light weight ejection seat which was fired by a 20 m.m. shell.

Hitler's German aircraft plants were being hit badly by American and R.A.F. bombers and so Heinkel's works in Vienna-Schmechat, Austria was the selected site of the prototype's construction. Here, also was housed Heinkel's engineering and design divisions.

Heinkel was awarded the competition design September 24, 1944. It was ordered into immediate production with a projected output of 1000 He-162's per month.

The German Air Ministry was well aware by late 1944 of the grimness of the war situation and Heinkel knew the difficulties facing him in the areas of material shortages, transport problems, and unskilled aircraft workers and more bombings. He faced them one at a time. He set up underground factories which were bomb proof. Since the He-162 was partially constructed of wood he employed German furniture and cabinet makers to short cut lead time. The wing, vertical rudders, landing gear doors, and nose cap were constructed entirely of wood.

The He-162 was the design creation of Heinkel's two top Aeronautical engineers: Herr. Gunter who was Chief Project Engineer, and Herr. Schwarzler was Chief Design Engineer. They and the other Heinkel engineers literally slept and ate at their desks in their design offices. Mattresses were placed next to their drafting boards to enable them to sleep a few hours at a time.

The final drawings were finished one day before the deadline set for October 30, 1944. These were rushed to the workshops.

The dimensions and characteristics of the He-162A:

Fuselage: A light all metal-monocoque structure with a wooden nose cap moulded to shape without machinery. Length—29 feet 8½ inches.

Wings: All wood construction including movable surfaces. Span 23 feet 7¾ inches, wing area 120 square feet.

Engine: One BMW .003 A-1 of 1760 pounds thrust.

Armament: Two MK-108 30 m.m. cannon with 50 rounds of ammunition per gun. The first production models used two MG-151 rapid firing 20 m.m. cannon.

Performance: Maximum speed 490 mph at sea level; 522 mph at 19,700 feet; 485 mph at 36,000 feet
Ceiling, 39,500 feet

Endurance (Full Throttle) 20 minutes at sea level; 57 minutes at 36,000 feet

Range, 165 miles at sea level; 267 miles at 19,700 feet; 410 miles at 36,000 feet

Weight: Normal—5,480 pounds
Maximum loaded—5,940 pounds

The landing gear was fully retractable. Jet fuel was housed internally.

On December 6, 1944, the prototype He-162 V-1 made its first flight from Schwechat airport near Vienna. The pilot was Captain Peter, Chief Test Pilot for Heinkel. The flight was made before company officials and the flight lasted for twelve minutes. Upon inspection of the aircraft after its first flight one landing gear door fitting had torn away due to defective bonding in the wooden door.

Not much attention was paid to this dangerous occurrence and another demonstration flight was scheduled for R.L.M. officials on December 10, 1944. Capt. Peter made a normal take off and was giving the large crowd that gathered a fine aerial demonstration of the He-162. When suddenly, upon his last low altitude run the right wing leading edge ripped away

continued on last page

Heinkel He 162 A-1 "SALAMANDER"

1. Sideview of Heinkel He 162A-1

2. Heinkel He 162A-1 Werknummer 120222

3. Heinkel **He-162 V-1**

4. Heinkel 162A-1 Werk Nr. 120086. First public showing of Hitler's secret jet plane, Hyde Park, London, October 1945.

5. Captured Heinkel He-162A-1 at Freeman Field Test Center, Seymour, Indiana. Note ballast weight atop nose.

6. Heinkel 162A T2-489 on exhibit at University of Kansas open house in June 1947.

7. Front view of He 162A-1

8. Heinkel 162A Werk Nr. 120077 on the flight line at Edwards Air Force Base during evaluation testing in April-May 1946. Note: gun parts were covered over for flight tests.

9. Heinkel He-162A Werk Nr. 220006 found at Munich, Germany by American G.I.s. T.A.I.C. crews had removed BMW .003 jet engine for tests. Note unusual duplication of aircraft serial number on rudder.

10. He 162A-1 Werk. Nr. 120227.

11. Abandoned Heinkel He-162A. This captured model had heavy armament of two MK-108 30mm cannon.

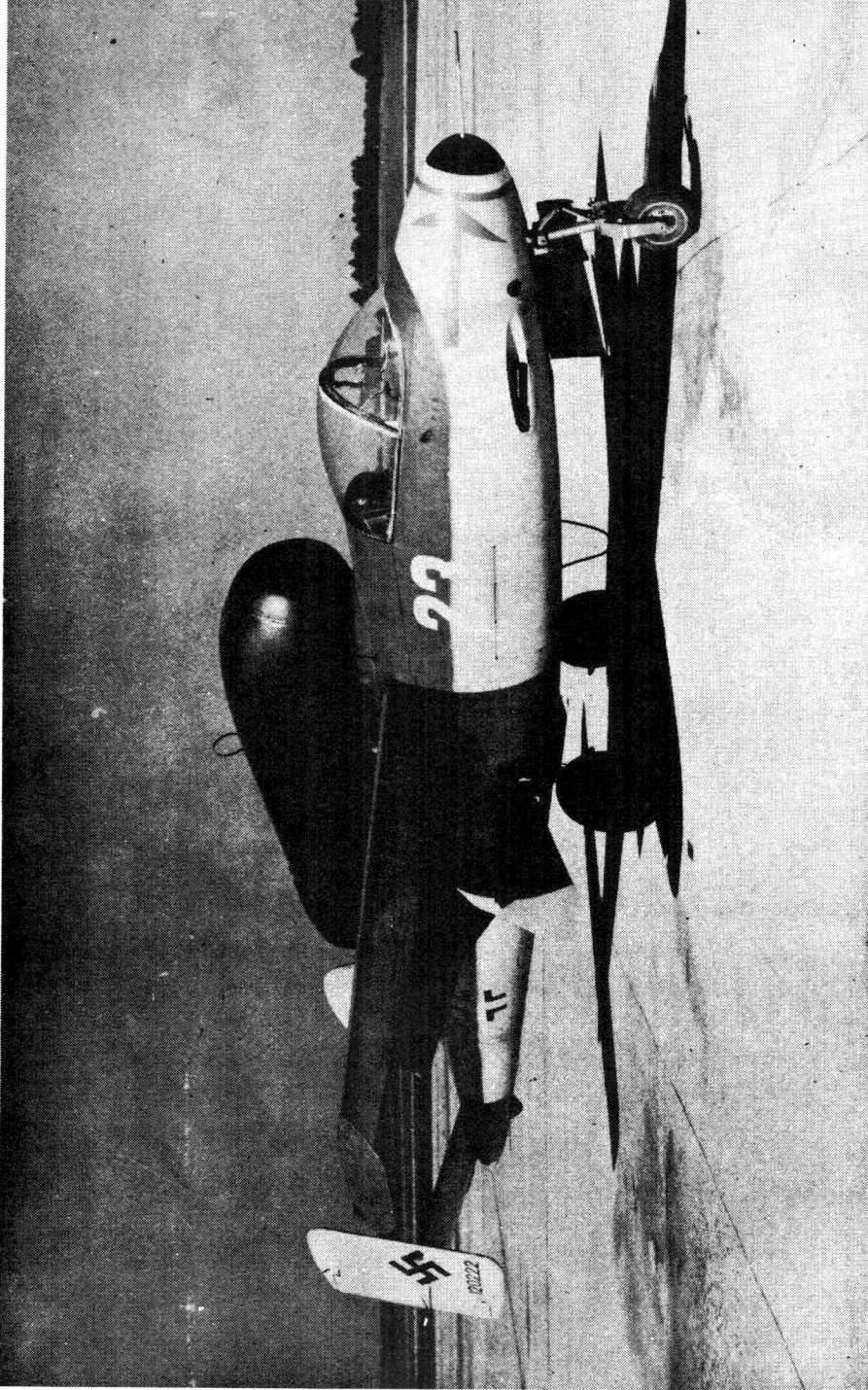

12. Heinkel He 162A-1

13. 14. Front view of He 162A at the Air Museum.

15. Canopy open. Note: "Nervenklau" is German word meaning "Nerve-Stealer."

16. Canopy and air intake.

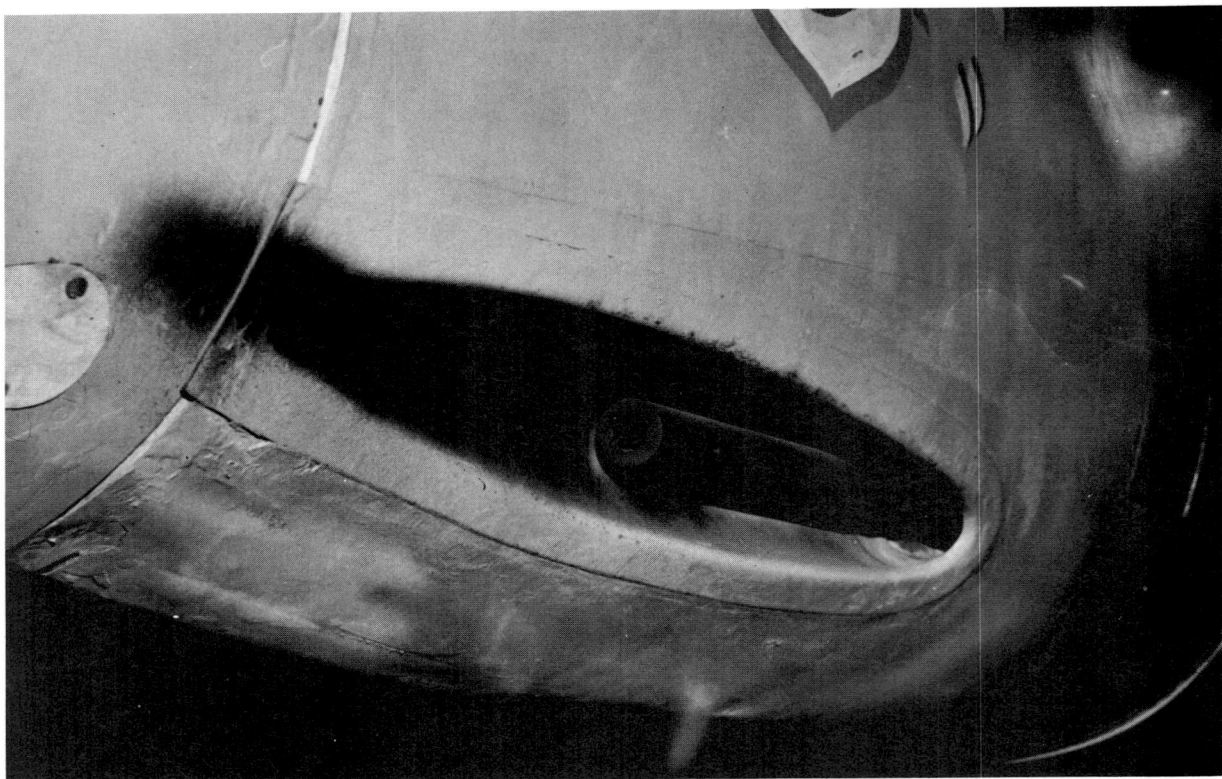

17. 20 mm MG 151/20 with 120 r.p.g.

18. Gun installation and blast tube. Note: electric wire leading to gunfire button on control stick.

19. Gun installation MG 151 20 mm.

20. Right side view of cockpit. In lower right corner note ejection seat grip handle and lever. Note American radio and oxygen installation in upper group.

Heinkel He 280 V-1

Engine: 2X Heinkel-Hirth HeS 8A Turbojets.
Speed: 578 m.p.h. at 19,685 ft.
Armament: 3 X 20 mm MG 151 cannon
Weight: Empty: 7092 lb.
Loaded: 9413 lb.

Heinkel He 280 V-1. The first twin jet fighter in the world. Flown on April 5, 1941 by Flugkapitan Warsitz before high ranking Luftwaffe officers in make-believe dog fight with a FW 190, it out-turned the FW completely. These two outstanding features were introduced for the first time: The nose landing gear and the ejection seat.

Heinkel He 162 B-1

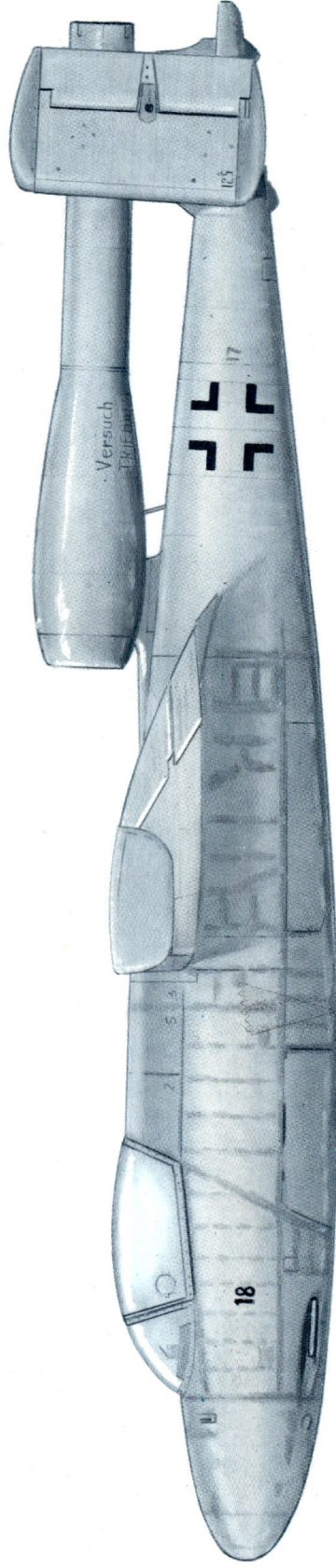

Heinkel He 162 B-1 with two Argus As 014 with 780 lb. thrust each, originally used on V-1. He 162 B-2 with one Argus As 044 impulse duct with 1100 lb. thrust.

Heinkel He 162 A-1 "SALAMANDER"

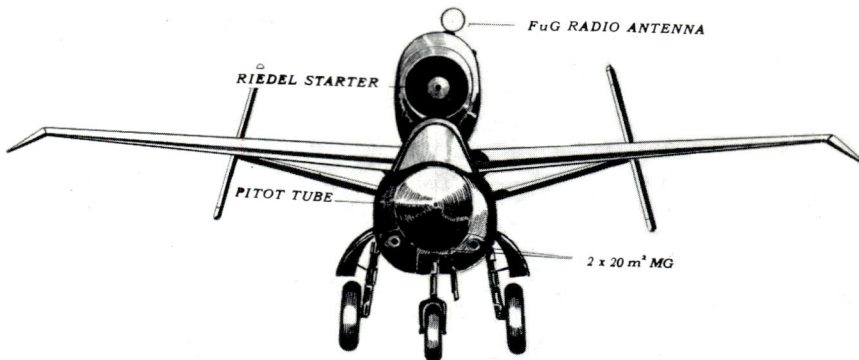

FuG RADIO ANTENNA

RIEDEL STARTER

PITOT TUBE

2 x 20 m² MG

Engine: 1 X 1760 lb. BMW 003 A-1 turbojet.
Speed: 522 m.p.h. at 19,700 ft.
Armament: 2 X 20 mm MG 151/20
Weight: Empty: 5480 lb.
             Loaded: 5940 lb.

AIR MUSEUM'S Heinkel He 162 A-1

The above color scheme is an exact copy of the He 162 A-1 displayed at the AIR MUSEUM.

Heinkel He 162 C

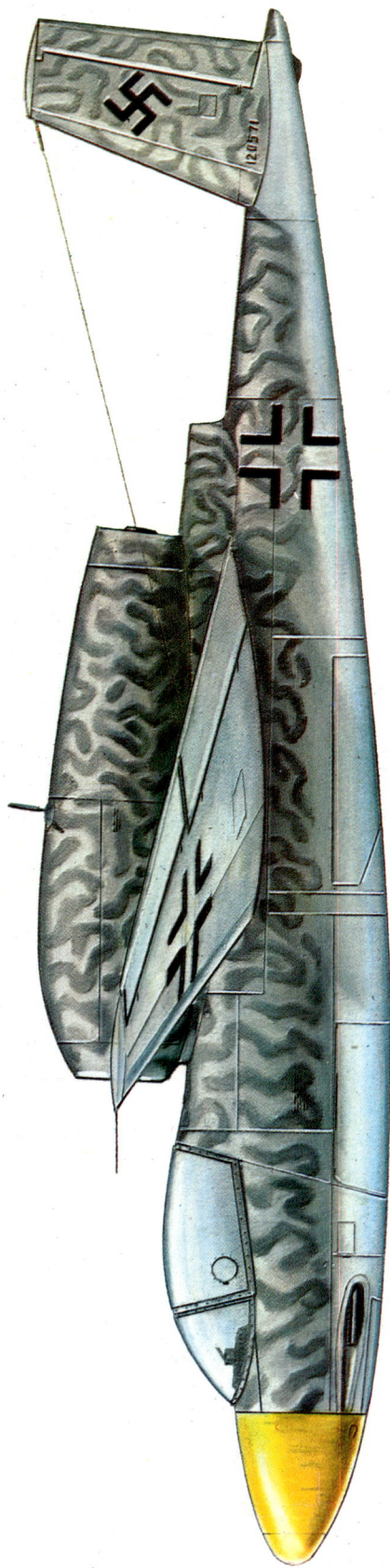

Heinkel He 162 C with HeS 011 2860 lb. thrust turbojet.

Heinkel He 162 D

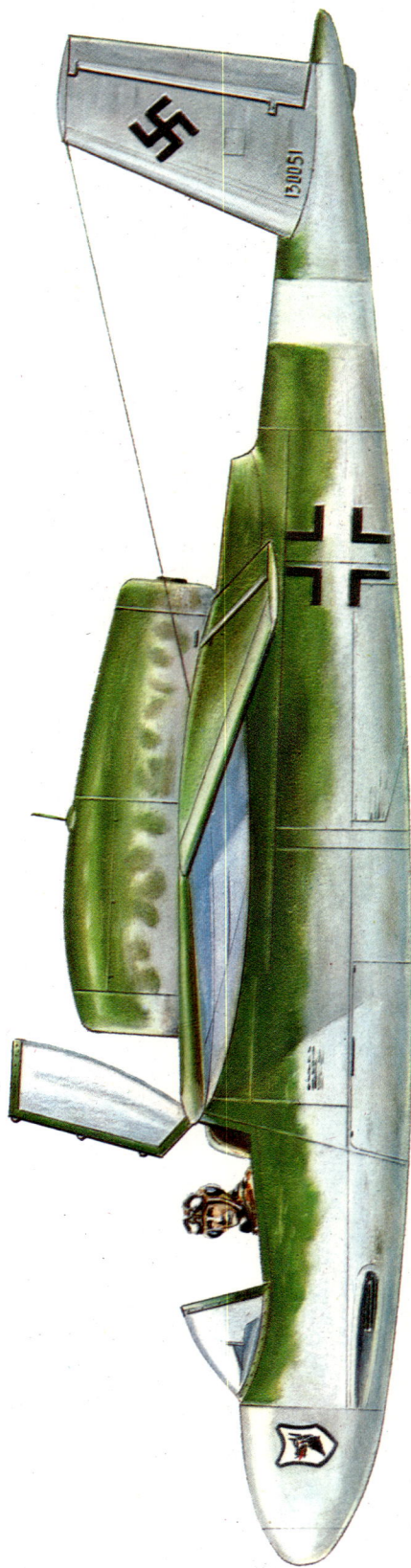

Heinkel He 162 D with HeS 011 2860 lb. thrust turbojet.

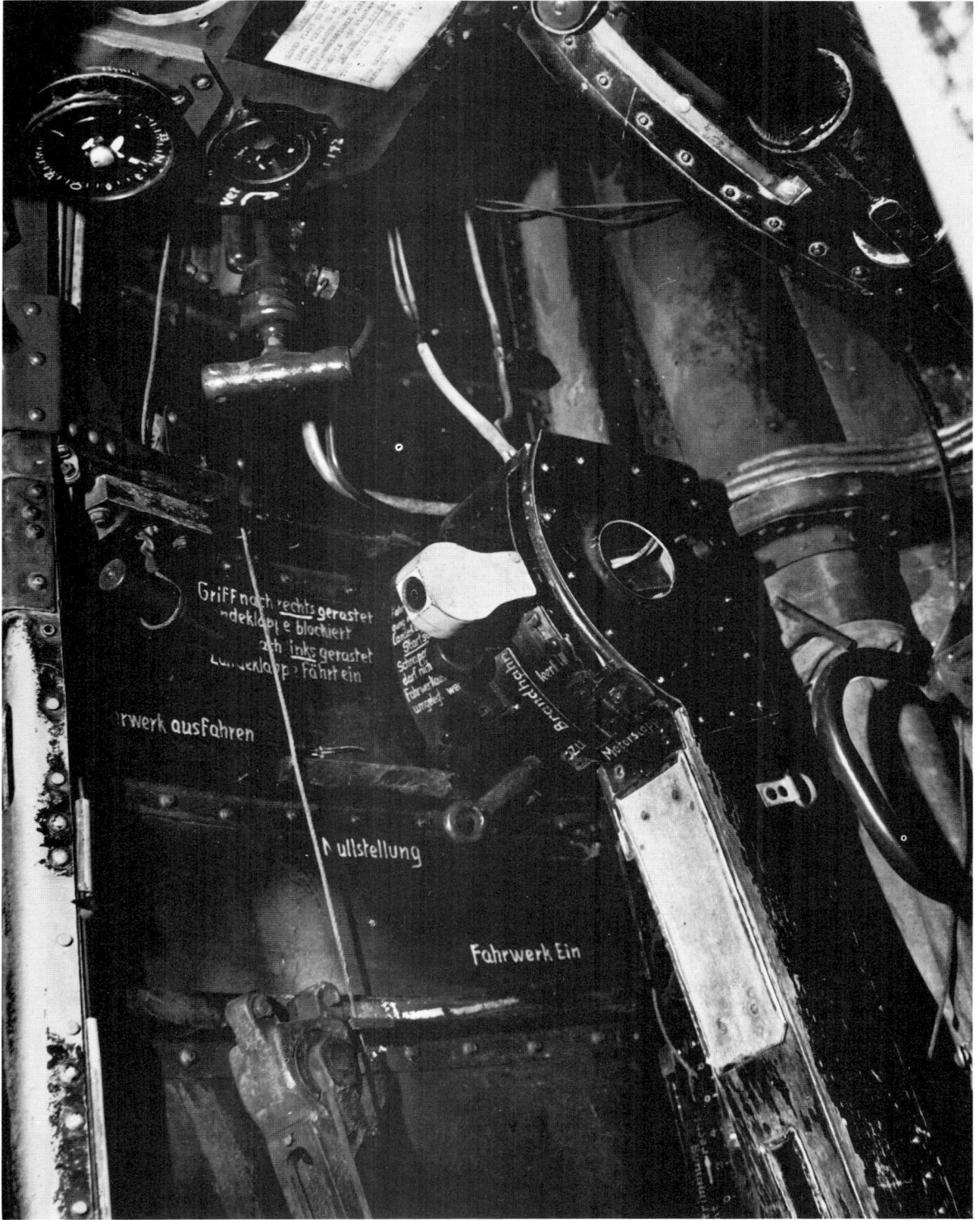

Griff nach rechts gerastet
ndeklappe blockiert
ach links gerastet
Landeklapp Fährt ein

rwerk ausfahren

ullstellung

Fahrwerk Ein

21. Left side view of cockpit. Note throttle control handle and below MG 151 20mm. gun blast tube.

Heinkel He 162

22. German Revi Gunsight

23. He-162A instrument panel
Note window panel below which allowed pilot to see when nose wheel was
retracted.

24. Nose of He 162, note pitot tube

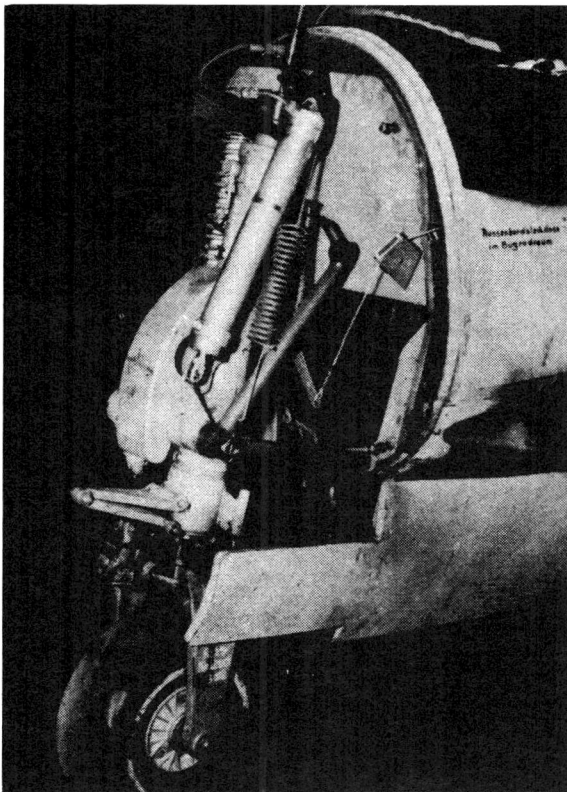

25. View of nose gear with nose cap removed.

26. 27. Nose gear

27 a. He 162A nose gear details close up.

28. Main landing gear assembly

29. Front view of left landing gear, tires are still the original German made "Continental Tires!"

30. Inside view of left landing gear, size 660 X 190 mm.

31. Close up of left landing gear.

32. Air intake

33

34. Aft cone of jet unit.

35. 36. Side views of BMW 003 A-1 German jet engine.

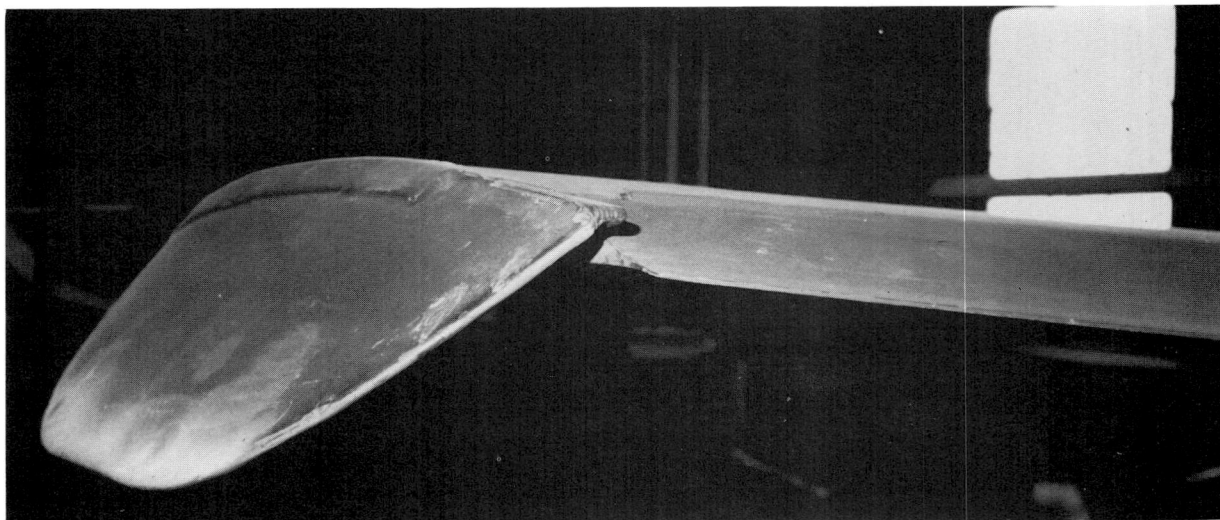

37. Wing tip showing negative dihedral.

38. Wing roots

39.

40. Empennage

41. Vertical rudder

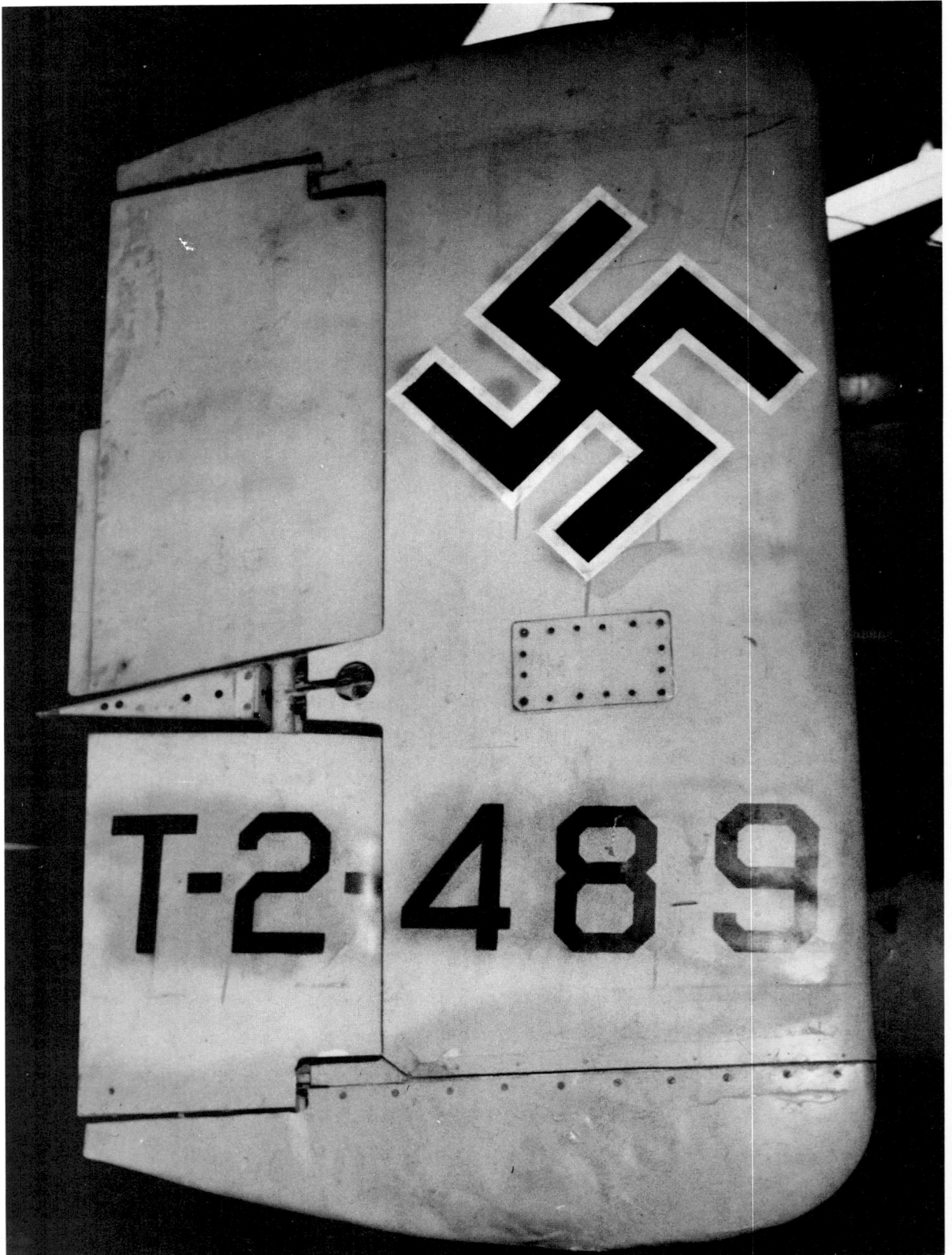

42. Close up of vertical rudder. Note: T2-489 was A.A.F. evaluation number.

CRACK UP OF PROTOTYPE HE-162

This dramatic sequel of photographs of a prototype aircraft is exceedingly rare. These photographs show the prototype He-162 V-1 as flown by Heinkel's chief test pilot, Flugkapitan Peter being demonstrated for high officials of the Luftwaffe and R.L.M. at Schwechat Air Base near Vienna, breaks up after a low altitude, high speed run, and was totally destroyed. Test pilot Peter was killed but development continued.

It would appear that the aircraft was pulled out of a dive too brisquely[1], the hi-"g" force first causing the right wing leading edge 4mm plywood skin wood structure to fail and break at the main spar[2]. Next came the right aileron, shortly after the pilot attempted to check the rolling-pitching motion that had arisen. Note that the right rudder and fin also began to disintegrate[3,4]. After a right roll came the final act as the aircraft flew apart[5,6,7].

The prototype was not equipped with pilot-seat ejection, so that the pilot had no chance to bale out.

In spite of the accident, the Heinkel 162 was put into production. However, only approximately 120 were built up to the time of the collapse of Germany. Factory drawings and half-finished aircraft fell into Russian army hands; eight complete He-162's came into Allied hands and a number were test flown following the war.

44. 45. Underground factories at Tarthun, Germany, **were** built to construct He-162A's when Allied bombers destroyed factories above ground. These captured He-162s were found intact on assembly line.

46. Rare line up view of Luftwaffe He 162's in Lech in the spring of 1944. Note: Squadron insignia of TG 84.

47. Heinkel He-162A taking off at Edwards Air Force Base. Test pilot—Major Bob Hoover, now customer relations director at North American Aviation.

48. Heinkel He-162A-1 Werk Nr. 120017. This aircraft was later scrapped.

49. Heinkel 162A Werk Nr. 120072 at Farnborough

50. He-162A-1 Werk Nr. 120222 on public display of captured Japanese and German aircraft following World War II. This aircraft is now in storage in the Smithsonian collection of the National Air Museum.

CAPTURED HEINKEL 162A AIRCRAFT

ENGLAND

Eight complete He-162 jet fighters were brought to England in 1945 by United States and British Technical Intelligence Teams. These aircraft came from the large pool of He-162's captured at the Luftwaffe jet air base at Leck. One of these eight was later given to Canada. The British Air Ministry assigned the following numbers to these aircraft:

R.A.F. Number AM 58   Werk Nr. 120021 Later given Number VH-526

R.A.F. Number AM 60   Werk Nr. 120086 First published showing of first captured He-162, Hyde Park, London 1945

R.A.F. Number AM 61   Werk Nr. 120072

R.A.F. Number AM 64   Werk Nr. 120097 Tested at Farnborough 1945

R.A.F. Number AM 65   Werk Nr. 120227 at R.A.F. college at Cranwell Sq. Aircraft Nr. 27

R.A.F. Number AM 66   Werk Nr. 120091

R.A.F. Number AM 67   Werk Nr. 530552 On display at R.A.F. Station, Colerne 1965

CANADA

Given to Canada by the R.A.F. this He-162A Werk Nr. 120076 has been repainted and was on display at the Toronto Museum until a short time ago when it was removed and placed in storage.

FRANCE

Was given a He-162A by United States Technical Intelligence crews for test evaluation purposes. It is now hanging up on display in the main building—Calais-Meudon Museum outside Paris.

AMERICA

Three complete He-162A aircraft were brought to the United States for test and evaluation. Two were sent to Freeman Field, Seymour, Indiana and one was shipped to Wright Field.

#1 Werk Nr. 120017 Sq. aircraft No. 7 on Fuselage Side. Was cut up for structural analysis
A.F. No. T2-494   at Wright Field.

#2 Werk Nr. 120222 Sq. aircraft No. 23 on Fuselage Side. Was flown at Freeman Field by
A.F. No. T2-504   Allied Pilots. Later taken to Wright Field and presented to Smithsonian Museum in 1947. Now in storage awaiting display.

#3 Werk Nr. 120077 Sq. aircraft No. 1. (Was originally No. 2 aircraft but has been painted
A.F. No. T2-489   over.) After preliminary showing at Freeman Field was shipped to Edwards Air Force Base and test flown in early 1946. Now on display at Ontario Air Museum, Ontario International Airport, Ontario, California. It is the only He-162A on public display in North America.

GERMANY

It is reported that one He-162A in the R.A.F. Air Ministry collection has been presented to the West German Government and it will be displayed in a German Museum in the near future.

Luftwaffe test pilot just leaving his Heinkel He 162A-1 after a successful at a secret air field in Eastern Germany. This He 162 was painted test flight in the Fall of 1944. He is here talking to engineering officers an overall light grey and the National insignia was solid black.

(Continued from Page 4)

owing to defective bonding. The He-162 V-1 then shed the right aileron and wing tip. It quickly rolled to the right several times and broke up. The test pilot, Peter, was killed. Scheduled work and production continued. The He-162 V-2 incorporatd anhedral wingtips to reduce the effective dehedral angle in the wing. Additional nose weight was added over the nose-wheel and the tail assembly was enlarged.

As experimental testing continued through the winter of 1944-45, pre-production aircraft and regular production aircraft were begun almost at the same time. The time interval of approximately two weeks separated each stage. When the He-162 V-1 first flew, quantity production was well underway. Ernst Heinkel used two factory labor shifts to accomplish He-162 mass production for a seven day week with no holiday leaves.

Other He-162 assembly plants were planned at Rostock, and Mittelwerke and by the Junkers plant at Dessau, Germany. A number of chalk mines were used especially at Mochling, near Vienna, Austria. Later it was planned to shift all final assembly activities directly to operational Luftwaffe airfields.

The all-metal fuselages were constructed at Barth, in Pomerania by Junkers at Ascher-sleber, and by Mittlewenke in an underground factory at Nordhausen where the German V-2 missile was also built.

The all wooden structures—wing, L. G. Doors, tail unit, and nose cap, were built with two production centers. One, at Erfurt in Thuringia, and another, near Stuttgart, Germany. This was done to take advantage of the large pool of skilled labor in the furniture and cabinet industries. Production problems were encountered but these were overcome. Additional wood parts were farmed out to many subcontractors.

Although the Heinkel He-162A never saw active combat there were several occasions when they were spotted by Allied Bombing crews. The author interviewed one ex-B24 Liberator crew member who well remembered the He-162 on a bombing mission over Austria. He stated during the He-162's run he was amazed at the speed of it as it passed through the formation. Other isolated sightings were made by Allied fighters, but contacts were not easily made as the He-162A could easily outspeed Allied propeller driven fighters.

United States and British Technical Intelligence crews were amazed to find large numbers of He-162's in production in German underground factories, as well as the large numbers of completed He-162's found.

It is of interest to note that the Heinkel He-162 was originally named "Spatz" (Sparrow) but it was later renamed "Salamander." However, Herr. Goebbels in the Propaganda Ministry utilized the "Volksjaeger" (People's Fighter) as a super weapon in his daily press releases, and this name became much more used than the former.

The He-162 marked a milestone in jet fighter aircraft design. Being the first high performance, lightweight jet fighter to be built with minimum materials and resources.